Katahdin Hair Sheep

Raising a Profitable Breed of Sheep

By
Darla Noble

Mendon Cottage Books

JD-Biz Publishing

All Rights Reserved.
No part of this publication may be reproduced in any form or by any means, including scanning, photocopying, or otherwise without prior written permission from JD-Biz Corp
Copyright © 2014. All Images by Darla Noble.

Table of Contents

Introduction

Chapter 1: What are Katahdin Hair Sheep

Chapter 2: The Katahdin Advantage

Chapter 3: Selecting Katahdin Sheep for your Flock

Chapter 4: Reaping the Benefits of the Katahdin Breed

Closing Remarks

Author Bio

Introduction

More than a few people would use words like 'dumb' and 'stupid' to describe sheep. Not me. I use words like 'trusting', 'habitual' and 'profitable'.

I use these words because they are what I know sheep to be. As someone who spent over twenty years as a shepherd to several hundred head of sheep, I can say in no uncertain terms that I know the potential for sheep when it comes to making your farm a profitable business.

The degree of profitability, however, depends upon the quality of your sheep. And when it comes to quality, you'll be hard-pressed to find a better breed of sheep than the Katahdin.

Katahdin are hair sheep; meaning they don't have wool. So yes, if your intentions for raising sheep are for spinning, you will need to choose a wool breed known for the quality of their wool. But if it's meat production you are after, look no further than the Katahdin.

Chapter 1: What are Katahdin Hair Sheep

Katahdin hair sheep are a breed of sheep that don't have wool but that do have a quality carcass that is significantly different than that of traditional wool breeds.

Wait! Developed? You mean Katahdins weren't on the ark? No, as a matter of fact, they weren't.

Katahdin sheep were developed by Matthew Piel, in the late 1950s and are named for Mount Katahdin in Maine; the location of Matthew Piel's farm. Piel was intrigued by the African hair sheep he saw in *National Geographic.* Were they really more prolific? Hardier? Did that hair coat really not require shearing?

The answers were yes, yes and yes; making Piel determined to add a few of them to his farm and try his hand at some genetic research to develop a breed of sheep that would exhibit all the positive qualities of the African sheep while adding the carcass size of larger wool breeds.

Over the next several years, the Piel Project crossed a number of different breeds of sheep in an attempt to develop a breed of sheep that:

*Had hair rather than wool

*Strong maternal instincts

*Would breed out of season; meaning they would breed/lamb any month of the year

*Were prolific—throwing twins and triplets consistently
*Had a meatier carcass

*More resistant to parasites and diseases than most other breeds of sheep

*Polled (no horns)

In addition to the "African" sheep (now known as St. Croix), Piel used Tunas, Suffolk, Cheviot, Southdown and Hampshire sheep.

The process of breeding and cross-breeding took several years, but the time and effort were well worth it.

Purebred Katahdin ewe and her triplet lambs

Today Katahdin hair sheep are a medium-sized sheep. They can range in colors from white or pinto to reddish-brown and even black—although black is usually the least desired color.

The hair coat of a Katahdin varies depending upon genetics. Hair coats are classified as A, B or C. The A coat is one that has no wooly fibers or that completely sheds all wooly fibers in the early spring. B coats are those that retain patches of wooly fiber over no more than 25% of their body. C coats are those that have more than 25% body coverage of wooly fibers that do not shed off each year. NOTE: C coated Katahdins are not eligible for registration.

Katahdins are also polled (no horns). Once in a while scur will surface—indicating a glitch in the genetics. This type of glitch should be a rare occasion at best.

The Katahdin offers a number of other specific traits that translate into advantages which will be covered in the next chapter.

In 1985 the Katahdin Hair Sheep International Association was 'born' taking its first members.

Today, KHSI http://www.countrylovin.com/KHSI/index.html is several hundred members strong and is recognized as one of the top ten animal registries in the world.

Katahdins are the most desired breed of meat sheep by both producers and consumers and continue to make a positive impact in the world of agriculture.

Katahdin ewe lambs

Chapter 2: The Katahdin Advantage

At this point you are probably doing one of two things:

1-Searching for the Katahdin breeder nearest you to purchase a starter flock

2-Telling yourself that the purpose of this book is to promote the breed, so of *course* I'm going to tell you how wonderful they are.

The purpose of this book isn't to sell you sheep. The purpose of this book is to educate and inform you as to the facts about Katahdins and why they are such a popular and profitable breed. You can think of this book as the "Dragnet" of sheep information—you know, 'the facts…just the facts'.

So without meaning to sound like a used car salesman, we're going to take a look at the FACTUAL and SUBSTANSIATED advantages of raising Katahdin hair sheep.

Katahdins are docile. By nature the Katahdin is an even-tempered, non-aggressive and gentle animal.

Now this isn't to say there are no ill-tempered Katahdins. I once had a large, red ewe we nicknamed Mary Poppins. The reason? She could fly! She was skittish, untrusting and at the first sign of being pushed toward a corral or trailer, she jumped up and over anything she could to get away. Needless to say, in a group o f several hundred ewes I didn't need her fly-in-the-ointment attitude, so she didn't stay in our flock for very long.
The same can be said for the rams. Male animals of any species are unpredictable and should be treated with respectfully cautious.

Katahdin ewes are docile and easy to work with

Katahdin sheep are excellent mothers. Several breeds of sheep simply don't make good mothers. Because I'm not in the business of bashing other breeds, I'm not going to name names, but there are some breeds of sheep that don't do well with more than one lamb. It's not that they can't raise twins—they just have trouble accepting both babies are hers. Once the ewe is made to take both of them, however, she will usually do okay.

Not so with the Katahdin. Katahdins are natural-born mothers. Twins and even triplets are the norm for most ewes—especially after their first lambing. The ewes instinctively clean the membrane from the lambs and begin nudging them to nurse almost immediately after birth. They also bond quickly. This fact was evidenced countless times by the way they weren't put off by my handling the lambs....

The ewes in our flock that lambed in the winter were given the luxury of spending about twenty-four hours in the barn in a lambing pen with their lambs. To get them there I would gather up the ewe's lamb(s) and go to the barn. First-time moms usually required me to walk backwards—keeping the lamb(s) out in front of me so they could see them. Most of the time, however, the ewes trotted along right beside me or even beat me to the barn.

The ewes are patient when new lambs don't always 'get it' right away when looking for nourishment. They watch over their lambs and always have a watchful eye on them if they wander off to play or explore their surroundings. And most of all, the Katahdin is highly efficient with her food; enabling her to produce enough milk to raise her lambs while maintaining her body condition.

Katahdin ewe just giving birth to twins

Katahdin sheep are prolific. While it is not unusual for first-time moms to have a single birth, each year afterward should result in twins. Triplets are not at all uncommon and can almost always be raised by the ewe, but must be monitored to make sure everyone is getting their fair share. Hey, there are only two teats, so what do you expect?

There have been ewes who have given birth to four and even five lambs; all of which lived thanks to the intervention of the shepherd willing to bottle feed one or two of them.

Katahdin sheep are low-maintenance sheep. In addition to the advantages we've talked about already, Katahdin sheep are considered a low-maintenance breed of sheep for the following reasons:

-Their resistance to parasites. Parasites (worms) can be a major problem for a shepherd. Parasite infestations can greatly affect the health, efficiency, productivity and ultimately, the profitability of any flock of sheep.

Group of Katahdin ewes on late summer pasture. Little or no worming is necessary when managed properly.

This is not to say Katahdins are immune to parasite problems. If nutrion is lacking and other management practices are substandard, parasites can wreak devastating havoc on any flock of sheep…including a flock of Katahdins.

-Their efficient use of food stuffs. Katahdins are extremely efficient with their feed. A Katahdin will put what they eat to good use; whether it be for meat, lamb development or body condition. They will stockpile standing forage by skipping over green fescue in favor of other grasses; saving the fescue for post-frost. The proof of their efficiency is seen in weight gains, meat analysis and body conditioning.

-Their adaptability. Katahdins flourish in any climate. Katahdins can be found in Canada, Mexico, South America and all over the United States.

NOTE: It is important to remember that low maintenance is not the same as no maintenance. Consistent and proper management simply means you will have considerably less labor-intensive work involved.

Katahdin meat is highly nutritional. The taste, texture and nutritional value of Katahdin meat is significantly better than that of other breeds of sheep as well as that of pork and beef and is nearly equivalent to chicken.

The absence of lanolin (due to the absence of wool) in their system alleviates the musky taste you find in sheep/lamb from other breeds. Instead, the taste is mild and goes well with a number of spices and seasonings. The texture is firm, but tender.

The nutritional value is eagerly accepted by anyone wishing to eat a healthy diet. The reasons for this are easily understood when you consider the following facts from studies done by the University of Missouri and University of Saskatchewan.

Cholesterol in a 100 gram serving
Chart is from the UofS study done in 1998

Katahdin Meat	44.4 mg
Domestic Lamb	72.5 mg

New Zealand Lamb	77.2 mg
Pork	69.0 mg
Turkey	72.0 mg
Beef	74.0 mg
Chicken	75.0 mg

The following information resulted from the 2001 analysis done by the University of Missouri

Katahdin per 100g serving:

Fat 2.5g   Protein 2.5g   Cholesterol 63mg

Other sheep per 100g serving

Fat 9.2g   Protein 20g   Cholesterol 88mg

Pork chop per 100g serving

Fat 8g   Protein 22.1g   Cholesterol 55mg

Chicken breast (skinned) per 100g serving

Fat 1.2g   Protein 23.1g   Cholesterol 58mg

When you combine these attributes of Katahdin meat, it should come as no surprise that that ethnic groups who consider lamb to be a dietary staple, prefer Katahdin. In fact, they will ask for it at the butcher shop and/or seek out Katahdin producers for the purpose of buying off the farm.

The social interactions between different ethnic groups and Caucasian Americans has led to an increased demand for lamb and they, too, are asking for Katahdin.

I stated at the beginning of this chapter and I will say it again…my intent is not to cut or bash other breeds of sheep, but to educate you on the advantages of raising Katahdin which in turn, result in increased profits to the shepherd.

Chapter 3: Selecting Katahdin Sheep for your Flock

Katahdins are like any other animal—some are better representatives of their breed than others. Some have better genetics than others which affect their overall condition and confirmation, their hair coat, prolificacy, how well they raise their lambs and how resistant they are to illnesses and maladies. Some are in a flock with better health management practices than others. Some are in a flock with better nutritional practices than others. In other words, some Katahdins are born lacking what it takes to be a top-notch animal while others may be perfect (or nearly perfect) but never get that chance to prove due to a lack of care and management on the part of the shepherd.

Katahdin ewe lambs.

The ewe lambs in the picture on the previous page are excellent representations of young, growing Katahdins. They are solid and thick-bodied with legs that sit squarely under their body.

It is important when selecting your first Katahdins, to choose carefully and to choose the best sheep you can find. Choosing the best you can find, however, does NOT mean choosing the most expensive. While it is often true that you get what you pay for, it is possible to purchase top-quality Katahdins for very a very economical price. It is also better to buy fewer sheep of greater quality than more sheep that are of lesser quality. Remember, you can build your flock by retaining your ewe lambs; allowing you to expand your flock while maintain the genetic standards of the adult ewes in your flock

Katahdin ewes and one to three week-old lambs

Selecting the best Katahdins for your flock comes down to one thing and one thing only—knowing what you are looking at and looking for.

So how does one go about knowing what they are looking at and looking for? By being familiar with the breed standards and knowing what constitutes excellent confirmation (body structure).

Six year-old Katahdin ewe shedding her winter coat in early spring. Notice her full-bodied frame in spite of just weaning twins and how her body and legs blend together nicely.

Katahdin Breed Standards & Confirmation

General Appearance: Medium-sized sheep with a hair coat devoid of wool except for possible wooly fibers in late fall/early winter that shed off each spring. Katahdins should have erect heads, bright, alert eyes and somewhat muscular legs that sit squarely on the four 'corners' of the body. Rams should be thick-muscled and masculine. Ewes should be strong, yet feminine in appearance.

Katahdin yearling ram. Notice his muscular body, thick neck and scruff. NOTE: He is shedding his winter coat.

Head: The head should be polled (free of horns/scurs), ears should be up, not drooping, eyes should be wide-set and the upper and lower teeth should meet evenly (no under or overbite). A ram's head should be much larger and thicker than a ewe's head. It will also most generally be less fine-lined.

Neck: The neck of a Katahdin should be strong, of medium length and should blend into the shoulder smoothly and evenly.

Shoulders: The shoulders should blend well with the neck and back. The shoulder of the ram should contain more muscle than that of the ewe. The shoulder blades should sit level or just slightly higher than the back and should be as wide as the middle of the back.

Yearling Katahdin ewes with excellent confirmation

Chest: The chest should be wide and deep. This is necessary for housing the heart and lungs. The chest of a ram should be equally as wide as the hind quarters. On ewes, it should be only slightly narrower than the hind quarters to give a feminine appearance.

Back: The back should be smooth, straight, broad and blend into the rump and tail area without any sharp drop-off. The loin should be long and wide, as well.

Ribs: The rib cage should be deep and broad to house the organs. The ribs should not protrude out from the rest of the body, giving a sunken appearance to the rest of the body.

Abdomen: The abdomen should be large and wide to house feed stuffs and have the capacity to allow the womb to expand when lambs are growing inside the ewe. It is acceptable for there to be a slight to moderate protrusion of the abdomen in well-conditioned sheep.

Rump: Wide, broad, thick and fleshy. It should be round and not drop off sharply in the tail area; blending into the top of the hind legs fully and smoothly. The rump of the ram will be more pronounced than that of the ewe.

Tail: Tail length is not an issue. Tail head width, however indicates the size of the loin. A wide, fat tail head is indicative of a larger loin than loins in sheep with skinny tail heads. NOTE: Docking is not necessary in Katahdins due to the lack of wool on their bodies.

Legs: Legs should be in proportionate size to the body in length and thickness. Back legs should be muscular and straight with properly angled hocks in the rear. Front legs should be strong and straight. The legs should always blend smoothly into the rest of the body. The feet and pasterns should be straight, as well and free of defect and disease.

Hair Coat: The hair coat of a Katahdin is graded into A B or C coated animals. Some will also include the grade of AA.

The AA coat is one that never has any wooly fibers at any time of the year. NOTE: Red Katahdins are more apt to be AA coated.

The A coat is one that displays very little wool in the fall/winter and sheds completely without evidence of any wooly fibers in early spring.

The B coat is one that will 'wool up' in the fall/winter but shed off at least 80% of the wooly fibers. The remaining fibers should only be found on the hind quarters, rump and/or underside.

The C coat is one that carries small areas of wooly fibers year-round or that sheds less than 50% of the fibers each spring.

NOTE: The wooly fibers are of no value, as they are hollow and cannot be used for making yarn or fabric.

A-coated Katahdin ewe—'wearing' little wool in January.

Scrotum and Udder: The ram should have two evenly-spaced and large testicles. The testicles should be tightly attached to the underside of the ram. The udder should be tightly attached to the underside of the ewe with two, even teats with a circumference not much larger than that of your little finger.

What to Avoid

General appearance: Sheep that are obviously lame, sway-backed, suffering from malnutrition, has a fouled tail and buttocks, has labored breathing or has a dull appearance in their coat or eyes should be avoided at all costs.

Sheep displaying any of these characteristics are indicative of a number of maladies including parasite infestations, Johnnes, pneumonia, foot rot and coccidia.

You can tell a lot about the sheep by looking around the farm on which they live. That is why it is always best to buy your Katahdins directly off the farm.

Head: A Katahdin's head should be sized in proportion to their body. The ram's head will be much larger and more square than a ewe's head. The ewe should have a head that is feminine and smaller, with a nose that is slightly to prominently more pointed.

They eyes should be clear and bright. Runny eyes and/or red eyes are signs of pink-eye; a highly contagious disease that has no cure and will be passed on to the lambs. Dull eyes can also be a sign of parasite infestation or fever. You would also be wise to look at the underneath side of their eyelids. The brighter red the underside of the eyelid is, the healthier the sheep. Paler pink and white indicate parasites are robbing the sheep of their blood supply.

The mouth should be free of an overbite or under bite. A malformed mouth prohibits proper chewing which can disrupt food intake and digestion. The Katahdins you select should also be free of blisters or wart-looking sores on and around the mouth/nose area. These blisters are evidence of sore mouth. Sore mouth is a highly contagious condition with no cure. While not fatal, or even serious, sore mouth is bothersome and unattractive. To prevent sore mouth you can keep your pasture free of abrasive plants.

Pasture free of thistles and other abrasive plants

Neck and shoulders: The Katahdin neck should be well-proportioned to the rest of the body. Do not choose a Katahdin for your flock that is long and thin, that does not blend well into the shoulders and is u-shaped. Likewise, stay away from sheep with narrow, pointy shoulders or shoulders. While the neck and shoulders of a ram will be much fuller and more muscular than that of a ewe, one that is significantly more muscular than the average well-conditioned ram should be chosen only if used on mature ewes, because even though Katahdins are bred for being easy-lambers, first-time moms may have trouble giving birth to lambs with the potential for above-average sized heads and shoulders.

Back: No Katahdin should be seen as a potential for your flock if the back (top line) is not straight and wide. The absence of a straight back indicates an overall weakness in structure, possible malnutrition and possibly even advanced age. Hey, sheep get old and worn-out, too.

A fallen or crooked back is also an indication of weakness in the legs or hocks. While some might consider this solely cosmetic, the truth of the matter is that posture and the strength of the back plays a major role in their ability to produce lambs that are strong and healthy as well as their ability to give birth quickly and easily.

As for the broadness of the back, avoid skinny sheep. That's about the only way to put it. Skinny sheep have a poorer carcass quality, they produce lambs with the same and they are generally less hardy and resilient.

Tail: Katahdin with tails that are fouled with manure indicate a problem. The problem may be as simple as being a bit of an over-eater o spring pasture, meaning the problem will easily correct itself in a day or so. On the other hand, however, the problem may be something more serious such as coccida, a parasite problem or Johnnes. So if the manure is fresh or if there is more than just a very small amount, steer clear. NOTE: sheep that are dry-lotted or living in a more confined area will sometimes have manure stains on their tails or buttocks which is the result of sitting or lying in bedding with manure on it.

The other tell-tale (sorry, I couldn't resist) sign of a Katahdin you want to avoid is one that has a skinny tail head. The fatter the tail head, the bigger the loin and the bigger the loin, the better the carcass quality.

Legs and feet: The legs and feet are the foundation your Katahdins will stand on …literally. That means it is essential for the legs and feet to be solid and sound. You will know this is not the case if the sheep you are looking at are limping, do not walk smoothly and evenly, do not stand squarely on all four legs, or if they stand looking as if they are about to sit back on their hind legs.

The hocks, which are the joint located near the center of the back leg, should only be slightly pointed out from the outside of the leg with no obvious bending at the joint on the inside of the leg. A sheep 'sitting' on their hocks, meaning hocks that are bent and weak, is a sheep without a good foundation.

The pastern is the heel part of the foot. It is that point at the back of the foot that should be sitting off the ground and not acting as part of the hoof.

Speaking of the hoof…NEVER buy a Katahdin with hooves that have any type of odor whatsoever. Yes, lift up their feet and examine the underside of their hooves. There may be dirt and grass and even manure, but when you scrape that away, you should have a solid foot pad and nail (hoof).

Sheep with long, over-grown hooves, but that are otherwise sound, should not be discounted. They may just be slightly over-fed or getting a bit too much protein in their diet. This causes their hooves to grow. Unless trimmed or given the opportunity to wear them down by walking on rocky ground, they will grow to the point of not being able to walk correctly. So if you do select a sheep for your flock with long hooves, be sure to trim them before turning them out into the field.

FYI: Kathadins with black hooves usually have less need of hoof trimming than those with white hooves. The black hooves are slower-growing and 'file' down easier on their own just by grazing.

Hair coat: The hair coat of a Katahdin has nothing to do with its ability to raise strong, healthy, meaty lambs. Nor does it have anything to do with its carcass quality, temperament or ability to breed. Hair coat is simply an indication of successfully breeding for breed standards. In fact, you can breed the wool off the lambs of most wool sheep by the third generation.

Most Katahdins have an A or B coat, however, so there is really no reason to buy those that don't. I'm simply saying that if there is a fluke in the genetic pool of your flock somewhere down the line and you end up with a C coat or two, don't cull them just because of that. If they are doing their job, leave them be.

Scrotum and Udder: Avoid rams with uneven testicles, testicles that are small and those that are not firmly attached to the body. The testicles should be free of sores, cuts, and knots.

The ewe's udder should also be firmly attached to the body and have only two teats that are uniform in size and free of warts, sores, malformation. The size of the teat is important. And no, bigger is not better. I once had a ewe that raised three sets of beautiful, healthy twins. But the following year, for some unknown reason, her teats were overly fat or large when she gave birth. They were literally so big the lambs could not fit them in their mouths. It was rough going for a few days, but eventually we were able to milk her and feed the lambs with a bottle. Needless to say we culled her before it came time to breed that group of ewes again.

IMPORTANT: While every Katahdin, no matter how old or young, big or small, should be in excellent health before adding them to your flock, it is necessary to point out that not all the breed standards will be evident in young lambs. This does NOT mean, however, that these lambs will not grow into phenomenal, or at least above-averages representations of the breed. For this reason, it is never wise to buy ewe lambs less than four to six months old that have been weaned for at least thirty days or a ram younger than a year old. Prior to that, you really have no idea what you are getting.

Two-day old lamb cannot be judged for breeding standards and confirmation yet

Chapter four: Reaping the Benefits of the Katahdin Breed

Yes, this book is all about Katahdins and why they are such an excellent choice for sheep producers literally all over the world. But the words would be just that—words—if there was no evidence to prove the truth of what I'm saying.

The proof I speak of is the continued demand for Katahdin meat. Ethnic groups seek out Katahdin producers to buy their lambs off the farm. Or if that is not possible, they request Katahdin meat from their local butcher.

There has also been a steady and significant rise in the number of Katahdins being produced over the past fifteen to twenty years. When we first started raising Katahdins in the mid-nineties, people were curious, couldn't pronounce the word Katahdin to save their life, thought they were goats and were skeptical as to what good they were without wool. Today if you mention the word Katahdin, someone will almost always say, "I know someone who raises those."

The fact that our country is once again becoming a more diverse society with various people groups mixing and mingling, lamb is becoming more popular with traditional white America. But…they prefer a mild-tasting lamb that is more palatable than older, muskier-tasting mutton or even wool breed lambs. The Katahdin fits the bill perfectly.

So for those who thought the Katahdin was a fad or novelty…think again. The Katahdin is here to stay.

The Katahdin is here to stay

Closing Remarks

The purpose of this book is not to be derogatory to other breeds of sheep and their producers. The purpose of this book is to highlight the main reasons Katahdins make sense—dollars and cents—for farmers everywhere.

Katahdins are a docile, prolific, hardy and resilient animal worthy of the positive attention they are getting.

Author Bio

Darla Noble is a native of mid-Missouri with over twenty-five years of experience as an author and ghostwriter. Darla's love of writing began in the fourth grade; after meeting up and coming children's author, Judy Blume, who, by the way, autographed Darla's copy of "Are you there, God...it's me, Margaret".

Darla's love for writing and family makes her work sought after in the Christian market, in the areas of parenting/family resources and inspirational nonfiction as well as working as a ghostwriter for educators and inspirational speakers. For more information about Darla as well as her other books, you can visit her website: www.dnoblewrites.webstarts.com.

Health Learning Series

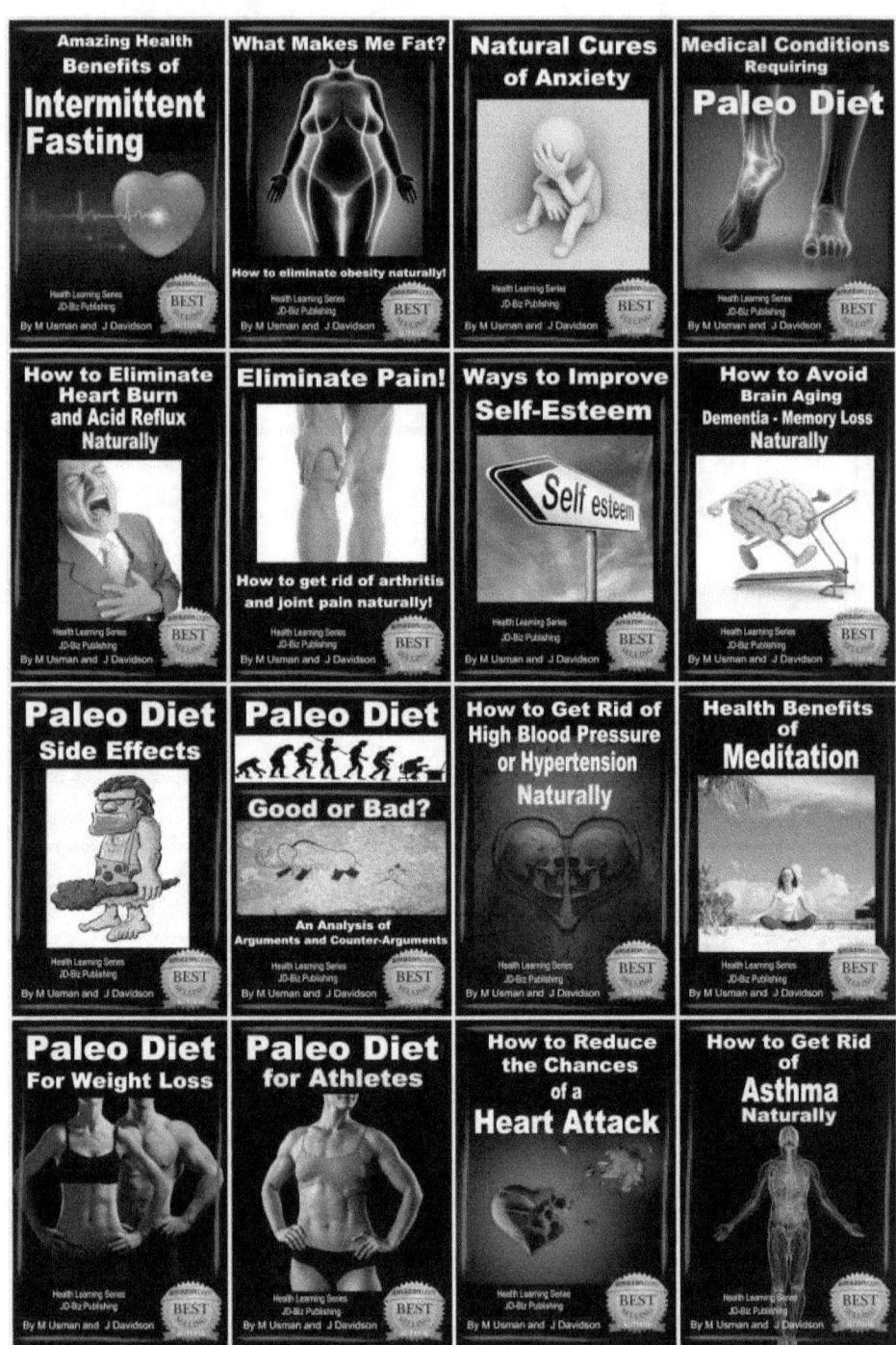

Amazing Animal Book Series

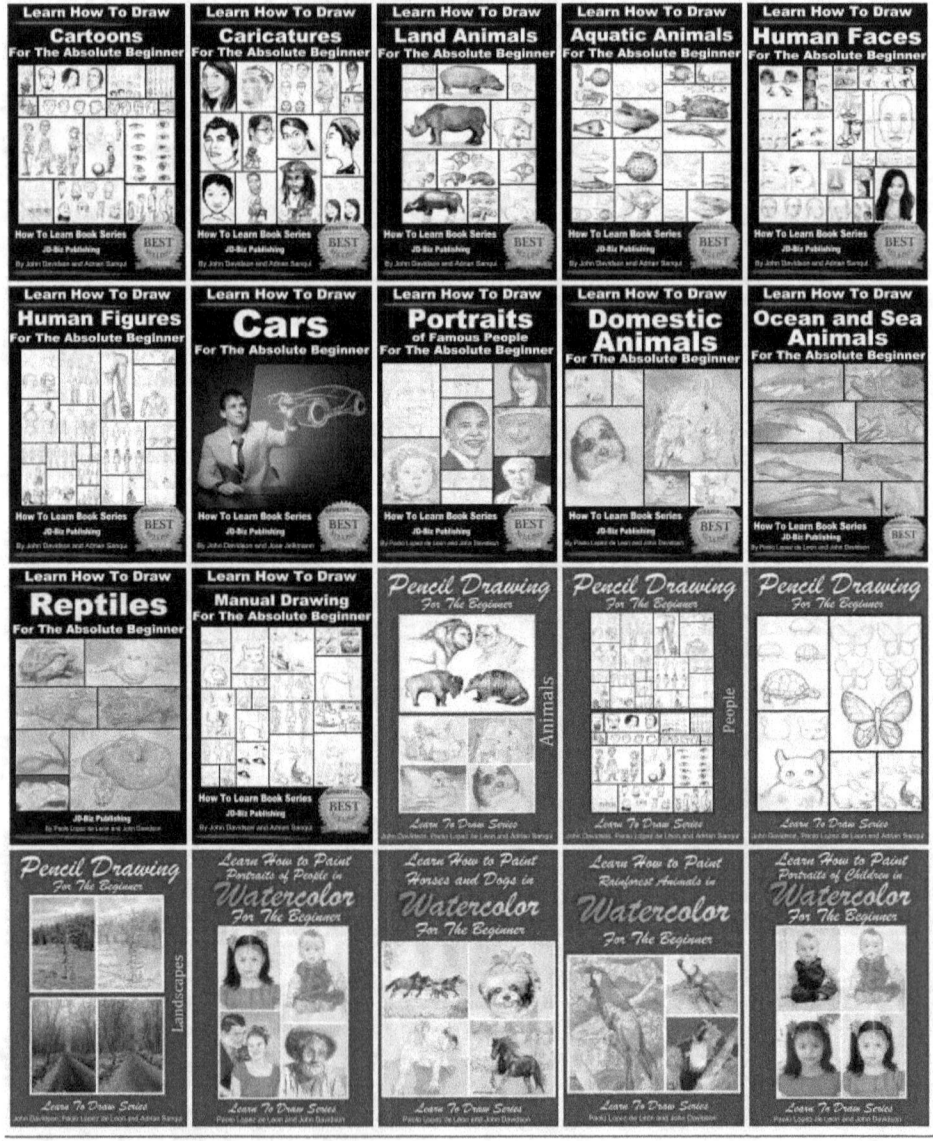

How to Build and Plan Books

Entrepreneur Book Series

This book is published by

JD-Biz Corp

P O Box 374

Mendon, Utah 84325

http://www.jd-biz.com/

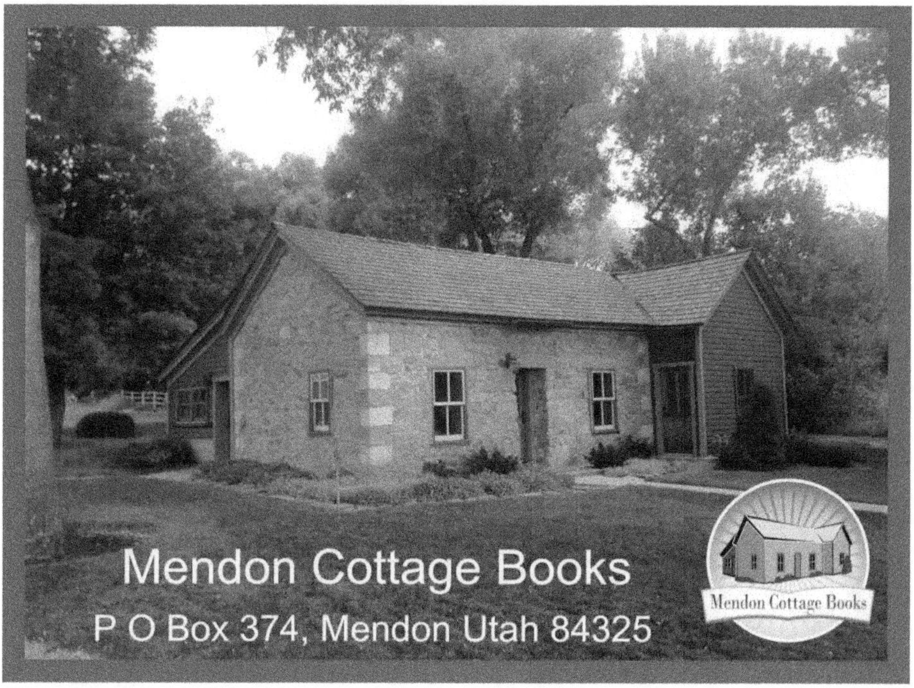

www.ingramcontent.com/pod-product-compliance
Lightning Source LLC
Chambersburg PA
CBHW070723180526
45167CB00004B/1597